STEVE TABER
ON BEEKEEPING

*Volume 2 from the archives of the Beekeepers Quarterly
for a new generation of beekeepers*

EDITED BY JOHN PHIPPS 2015

PHOTO
STEVE TABER (by John Phipps)

First published by Northern Bee Books, 2015.
First Edition.

NORTHERN BEE BOOKS,
Scout Bottom Farm,
Mytholmroyd,
Hebden Bridge,
West Yorkshire HX7 5JS.

www.northernbeebooks.co.uk

ISBN 978-1-908904-88-1

Printed & Bound by
Lightning Source UK Ltd.
www.lightningsource.com

CONTENTS

Foreword By John Phipps 7

Editor's Note 9

1. WHAT DO BEES WANT? 11

   BKQ 25 Spring 1991
   Ask your Bees a Question 11

   BKQ 36 Winter 1993/4
   Arguments about Bee Hives 15

   BKQ 69 Spring 2002
   Do Your Bees Have Enough Room? 19

2. HEALTH AND PESTICIDES 24

   BKQ 38 Summer 1994
   Artificial Status of Bee Health 24

   BKQ 40 Winter 1994/5
   The Origin of Pesticide Application by Plane 27

3. QUEEN BEES AND BREEDING 30

   BKQ 53 Spring 1998
   Queen Rearing without Grafting 30

BKQ 86 November 2006
The Right Bees for Queen Rearing 33

BKQ 49 Spring 1997
Multiple Mating of Honeybee Queens 35

BKQ 75 November 2003
Finding Queens 39
Finding Queens in Nucleus Hives 39
Colony with Just One Box for Brood 41
Colony With More Than One Box For Brood 42
Finding the Queen in a Bad-Tempered Colony 42

BKQ 37 Spring 1994
Breeding Bees in Villebrumier 45

BKQ 74 July 2003
Why Keep the British Bee? Why Indeed! 49

BKQ 47 Autumn 1996
Brother Adam and the Buckfast Bee 50

4. HONEYBEE BIOLOGY, BEHAVIOUR AND EXPERIMENTS 55

BKQ 44 Winter 1995/6
Bee Eggs & Larvae 55

BKQ 45 Spring 1996
Bee Larvae 59

BKQ 57 Spring 1999
Drones 61

BKQ 58 Summer 1999
Drones Again 65

CONTENTS

    BKQ 52 Winter 1998
        Naive Bees 69

5.   HONEY 74

    BKQ 61 Summer 2000
        Cream your Honey 74

FOREWORD.
By John Phipps, Editor, June 2015

Since its first publication in 1984, The Beekeepers Quarterly has attracted some of the best writers from all parts of the world. Many of the contributors have had a specific interest in a particular sphere of beekeeping and have brought expert knowledge to our pages, thus keeping our readers up-to-date with the latest advances in beekeeping science, technology and practice.

Most of our writers have had many, many years of beekeeping experience behind them which has always allowed them to see emerging problems in the context of past events, giving them an overview which cannot be surpassed by recent writers.

Of importance, too, is the fact that their perspective is not only historical, but also global, for they have concerned themselves with what is happening on an international scale which has given them a decided advantage over those who have a more parochial interest in the craft.

The Beekeepers Quarterly has always been concerned to present aspects of beekeeping in the widest possible way and this has been reflected in the work of our team of writers, many of whom who have been connected with the magazine for decades.

Much of the material written over the years is still of great value and relevance to beekeepers who have recently started out in beekeeping. So, in order to make this material available, we are publishing volumes of past articles written by our major contributors. Inevitably, there are areas where a very minor amount of the information has changed - this is particularly so regarding beekeeping pests and their treatments and the reader is therefore encouraged to seek the latest advice on recommended treatments.

I met Steve Taber on just one occasion. It was immediately obvious to me that this gruff-looking, larger-than-life beekeeper had a powerful personality, and that although opinionated and provocative in his writing, his knowledge of beekeeping was the result of many years spent both in commercial apiaries and research laboratories where he questioned the legitimacy of many beekeeping 'truths' through carrying out experiment after experiment, until he was satisfied with their outcome. In his articles Steve encouraged beekeepers to question what they did, why they did it, and to learn about beekeeping by spending more time at the hive side studying their colonies.

EDITOR'S NOTE.

Steve Taber was well-known throughout the world, especially in America where he lived and worked for most of his life and contributed regularly to American beekeeping journals. He began beekeeping as a hobbyist in Columbia in 1938 and turned to commercial beekeeping 3 years later in New York State with a man who operated 1800 hives. After a brief spell as a flier in the Navy during the war he attended Wisconsin University and studied bees with Dr. C.L. Farrar. During this time he worked bees in Ohio on the USDA's Kelly Island hybrid bee project. In 1950 he was appointed to the Baton Rouge, Louisiana USDA Lab. working with Dr. O. Mackenson and remained there until 1965 when the government transferred him to Tucson, Arizona. Steve retired from U.S. Dept. of Agriculture in 1969 and started two bee businesses in California - Taber Apiaries and Honey Bee Genetics. Here his time was devoted to breeding bees with special characteristics, primarily resistance to AFB, chalkbrood and *Acarapis woodii*. He then moved to the South of France, near Toulouse, working with the bee breeder John Kefuss and also concentrated on chalkbrood research working with Dr. Martha Gilliam at the Tucson Bee Lab.

SECTION 1.
WHAT DO BEES WANT?

BKQ 25
Spring 1991

Ask your Bees a Question

Most beekeepers are keeping their bees because they like to get a bit closer to nature. There are, of course, several of us who want to make it a source of income too, but even here, most commercial beekeepers are basically nature lovers. The intriguing thing to me then, if all of us are so keen on nature, why is it that we abolish a large part of nature in the bee hive?

Well, maybe you don't, but I'll bet you do. In the past 20 years or so we have been deluged by various manufacturers of plastic combs and foundation for the bees to build their comb onto. This is my beef, or my complaint, mostly because I don't see any need of it or for it. And they are terribly expensive. But, yes, hobby beekeepers just love to spend money on their stupid hobbies. I mean, why have a hobby unless you can spend your money.

Did you ever bother to think about what the bees would prefer to have in their house (hive) if they could tell you? Well, sometimes why not ask them to find out? I have been asking that question of bees for many, many years. The only problem is you have to be able to speak the bees' language, they refuse to learn English or even French for that matter. So when you start talking to the bees, or let's say communicating with them, you have to be very careful not to introduce any artefacts that will confuse the little darlings.

A very easy first type question to ask them is "Hey you guys, which of all these flowers do you prefer to collect nectar from?" They will tell you the answer pretty quickly as you can readily observe by taking a walk along any garden path and seeing bees all over apple blossoms, dandelions and clover. But not so on most pear varieties or the horticultural roses.

Now that you are in the proper frame of mind, let's go into the bee hive itself. I asked the bees this question "Hey, if you were going to live inside a box where would you put your nest?" To have the bees answer that question, I put them in a box that was a 4 foot cube, some 30 or so, over a period of 2 years. Their answer was, "It doesn't make any difference, except, not on the floor". Well then, I asked the bees another question, "If you are going to build your nest would you prefer a warm or cool place?"

To get their answer, I had built for me a long box, about 8 feet, cooled at one end and warmed at the other, and I then dumped swarms into the box at various places for them to build their comb. Their answer was, "It doesn't make any difference."

These were experiments that I made when employed by the US Government some years ago. The reason I bring them up now for your thoughts, is that, beekeepers, I am talking to YOU, now rarely ever bother to ask their bees questions, they want some government experts or university people to do all of that now. It was not this way a hundred years ago, then plain old observant beekeepers were discovering all of the things we call facts today.

Langstroth was a religious preacher, Root was a jeweller and Miller, one of America's first commercial beekeepers from Illinois, was a medical doctor. There is today a great lack of scientific people who spend their time watching, observing what goes on with their bees. I call this "rocking chair" bee research. When I first started my bee research career back in 1950, I was interested in the question of "Hey, you virgin queens, when you fly up and away, how many husbands do you take?"

Partially to answer that question, I sat in an old rocking chair in the apiary with a bunch of virgins in mating nuclei watching to see when they flew, how long, and whether they came back with a mating sign or not. Have you ever thought of taking an old rocking chair (Pardon me, but I don't know whether you people in England use rocking chairs or not, I am just a stupid American, a colonist, remember?) out in the apiary to watch, and I mean WATCH, your bees.

Of your various stocks, do some get up earlier in the morning than others? When it gets a bit cloudy, do some work and others quit? You can see that some produce more honey than others, but what about disease, AFB, or chalkbrood, to make it more simple, how about their bottom boards, are some very dirty and others clean? You know all these things are genetically controlled and in any apiary you should see many different variations in the behaviour of your bees.

Now then, I want to come back to the original preface about the combs in the hive and my aversion to plastic ones particularly. About 25 years ago, perhaps even more, I began to think about asking the bees the question, "Hey guys, why do you want to build drone comb sometimes and worker comb other times?"

This question was prompted by the following observation, I had this colony during a honey flow building comb in frames with no foundation, you know, no wax placed there to direct their activities. The hive contained 10 frames. I would remove one on the right and give an empty one on the left. Five days later I would do the same. And every five days remove the comb on the right, give an empty one on the left.

At first the bees built solid drone cells, top to bottom on each and every new frame introduced. Then all of a sudden, as if by magic, the bees decided they had enough drone comb and built nothing but worker size cells. For the first time in my career I saw the bees change from building drone cells to worker cells on a comb. I began to think, "Why do bees decide to build drone comb anyway?" or "Why do bees build worker comb?" You know something, that is a hell of a question to ask the little darlings to answer and so far as I know there is no answer yet.

But I keep drifting off my subject, all my English teachers said, "Prepare a good outline for your article first, Steve", and yes, I usually got very poor grades in my native tongue. When you have your bees build comb, why do you have them build on what is called "foundation" made of beeswax or plastic? Have you ever thought of having the bees just build their own comb?

This great fundamental thought came to me some 25 years ago, which was "Let the bees decide what kind of comb to build", I will decide only where they will build it. To implement this I started to install just empty frames into my hives. Presto, combs were built and when I would examine the combs the comb would fall out. Reading Langstroth, (you know there is no better way to learn something than

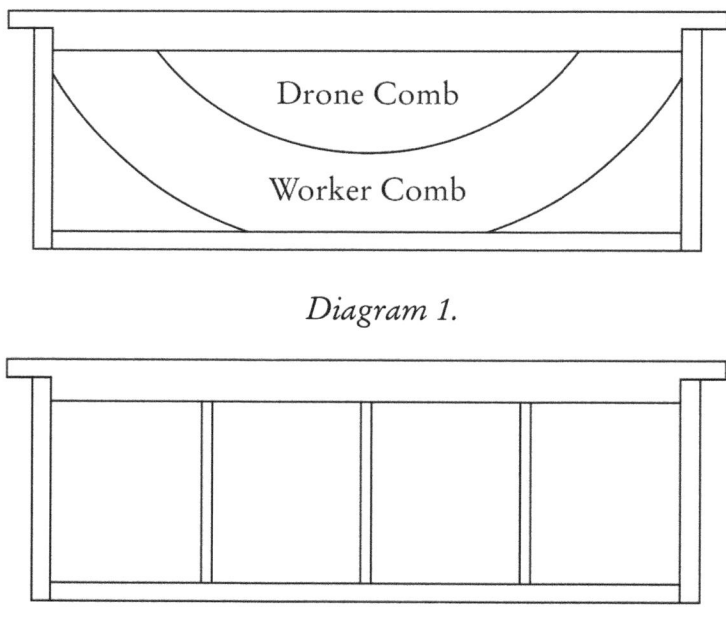

Diagram 1.

Diagram 2.

to read) I found that he placed small sticks inside the frame which the bees built comb on, that would prevent the comb from falling from the frame when it was being manipulated.

So I began boring holes top and bottom of the wood frame to insert and glue in place, wood dowels, any size, it makes no difference to the bees. I use 3/16", and three of them in each frame. I asked myself, "Why should I spend money to buy this expensive foundation when the bees will do it themselves?" Yes they will. Will they build straight comb? Hell no. When the bees don't do it right, tear it out and tell them to do it over again. Or if you want, you can tear it out and stick it back where it should have been put by the bees, but wasn't, and hold it in place with a rubber band so they can stick it in place.

You see, what I have been advocating to you is getting nature back into your hobby or vocation. Sure, you will get much more drone cells built by the bees but who has proved or shown that this by itself causes less honey production? Why not ask the bees this question? And somebody, somewhere should be able to figure out how to ask the question to the bees, "Why do you guys want to build drone comb?"

BKQ 36
Winter 1993/4

Arguments about Bee Hives

"If the bees could talk and you asked them what would be an ideal size and shape of a bee hive, what would they say?"

I enjoy being contentious, disruptive and argumentative and two articles in the Autumn '93 Beekeepers Quarterly gave me an excuse to do just that. They were "Make a cottage hive" and "What's the best sort of hive?". This brings to mind all sorts of problems which I wrestled with in years past doing research on bees while employed by the US Dept. of Agriculture at Tucson, Arizona, the site of a very large bee research institute. At that I time I was a friend and co-worker to a very knowledgeable agricultural engineer, C. Owens, assigned to solve beekeeper problems there.

At the time, and for many years previously, I had been under the impression, given to me by my University Professor, Dr. C. L. Farrar, that any size and shape of box which provided sufficient protection from severe elements of nature was the correct size and shape of a bee hive provided it had movable frames. Well Mr. Owens attacked that idea over and over again. He said and said many times until I finally listened. "If the bees could talk and you asked them what would be an ideal size and shape of a bee hive, what would they say?" Of course he would re-phrase the question months later and I suppose it took me a year before I thought, why not let's just ask the bees that question.

I became pretty good at asking the bees questions and having them give me straight answers, but this is not an easy task because they don't talk English or for that matter American English or, in fact, any other of the many different human languages.

What has to be done is to phrase the question in the words of an experiment which has to be designed to remove, as much as possible, human bias and prejudice. Our experiment was designed as follows: we built boxes 4' in all dimensions, a cube. On one side, 2' from the bottom and from the side, we placed an entrance hole. In the centre of the box we placed a stick, a dowel, 3' high with a small table attached to it. Sixteen of such boxes were built and we conducted this experiment for two years giving us 32 observations. Since some bee swarms start with different quantities of bees, we did too. Some were 2 pounds, some 4 and some 6.

The boxes were made so that either of the two opposing sides could be opened for an observation. At the end of three weeks of egg laying by the queen and before any new bees emerged, the colonies were killed and many different measurements and observations were made on the construction of the comb and arrangement of honey and brood.

To sum up what most beekeepers think is the most important information, and the reason for my writing this article, is that in all 32 instances the brood nest was exactly spherical, a round ball not a squashed-shaped ball which exists in most man-made bee hives. (Information is published in the journal, Animal Behaviour 18:625-6321970, entitled: Colony founding and initial nest design of honey bees.)

We then constructed hives with frames exactly 18" square made from steel and placed them in a box which would hold 13, spaced exactly 3/8" apart. Six of these were made and we got brood nests that were exactly spherical and resulting bee populations that were enormous. However the Tucson area is not a good honey producing one, very frequently, so no honey production records are available. We decided the brood box was too large and another six were made where the frames were rectangular, measuring 18" x 9". More honey was put into supers with this hive than the first one. These two hives were described in Gleanings in Bee Culture in 1974 in the January issue.

An interesting feature of all 12 of these hives is that on the top, sides and bottom were placed two round 3/8" rods to provide proper bee spacing. On the middle of each side was placed a pipe flange where a six inch piece of pipe could be attached. The hive was manipulated by a person on each side picking up on the small piece of pipe and turning the hive 180 degrees, a cork 3" in diameter was removed from a hole in the now front for the bees entrance and placed in the rear hole in what had been the entrance. Now you have to admit that is a cute manipulation technique and it works beautifully too.

To come back to the beginning, building a "Cottage Hive" is a grand idea but not enough attention is given in the article to the necessary bee space which exists between all parts. The 3/8th" is a good average, but you can use 1/2" too if you prefer, the bees use both. And in fact in honey supers the bees use 1/4".

I heartily agree that beekeepers should get out into their wood shop and make up some of their own equipment. I fabricate my own queen mating boxes but I make them very simple, and none are painted white. I think that is a terrible colour for a bee hive. For two reasons: if you use different colours it gives bees a better chance of not getting lost and going into a hive other than their own. Secondly, the many different colours of my hives makes for a very attractive aspect to my garden. In fact if you want to really get fancy with your painting, paint pictures and landscapes above the hive entrances as the old timers did many years ago. Some of those old hives which I have seen are really beautiful. But, if you live boxed in with neighbours close by, remember nobody loves bees but a beekeeper and in fact all the rest of the world thinks the world would be better off with no stinging insects in it. If that is your case, hide your bees and use a paint colour that will camouflage and hide them. What your neighbours don't know and don't see will keep them from complaining.

The second article "What's the best hive"? unfortunately contains some mis-information. As far as I know, no one in the world today uses a "Langstroth Hive" as described by the Rev. L. L. Langstroth. Some years ago a carpenter in the wood shop at the Tucson Bee Institute, Louis Gasca, built a copy of Langstroth's hive and it was taken to the Apimondia Congress in Moscow where it won a blue ribbon. All hives all over the world that have movable combs, spaced from 1/4" to 1/2" apart hanging by their top bars from a support are Langstroth, or to be more accurate, modifications of Langstroth's dimensions.

As to the perfect size box to hold these frames it is which ever provides a circular or spherical brood nest during all beekeeping seasons without crowding the bees so that they swarm.

What do I use and why? I want all my boxes and combs to be interchangeable and not so heavy that I have trouble lifting them. It is much more important to me that a full super of honey does no harm to me when I lift it as I already have a damaged spinal disc, that means not much more than 40 lbs of weight. And since I have graduated into the ranks of certified, guaranteed OLD MAN, (I will be 70 in a few months) most of my muscles and joints ain't what they used to be. The box size I use then is 6 5/8" deep holding 10 combs. After combs have been built I reduce their number to 9 making removal quicker and easier. In the USA this size of equipment goes by two different names, "Illinois Shallow" and "Modified Dadant Shallow". If you are an elderly beekeeper or one who's muscular strength is on the low side, the perfect size hive for you is the one that weighs the least and the bees will do all right.

BKQ 69
Summer 2002

Do Your Bees Have Enough Room?

We had our summer meeting in July at Clemson, South Carolina. Amongst the guest speakers was Dr James Tew, one of the most charismatic speakers about bees it has been my pleasure to know. One of the talks (he gave several) was about swarming and its control. What amazed me about his talk was the fact he was keeping most of the colonies in either one box, a standard 10 frame Langstroth, or two.

It is a lot of work to keep bees in two boxes and manipulate them so the queen can develop a maximum population. And I am sure that most of you have noticed, when you do keep bees in this manner, that when you put on supers if you don't put on a queen excluder the queen is immediately up there enlarging her brood nest. In order to stop that it is essential to put an excluder over the top brood box.

Now stop and think a bit. If the queen is moving out of the two brood boxes then she thinks, or feels, or knows that she does not have enough room to suit her egg laying potential.

She is trying to tell you that she needs more room.

Let's go back to the problem of swarming for a bit. I spent two summers working for commercial beekeepers in 1941 and 1942. They were each operating about 2000 colonies for honey in New York State. They both spent most of the month of May and early June going through the brood nest and destroying queen cells. One of these men would "Demaree" a colony if it was bound and determined to swarm. They were both spending a lot of time and money cutting queen cells - an operation which could have been avoided.

The Demaree system of swarm control was developed over a hundred years ago and it stops swarming, I guarantee it. It also severely disturbs the colony and the queen's egg laying for about ten days. The system, briefly described, is as follows:

Find the queen and put her in the bottom box with mainly empty combs, plus two frames of sealed brood and ensure that all queen cells have been destroyed.

Put a queen excluder on top of the bottom box and then add at least two supers.

Place the rest of the colony with all the remaining brood in the top box, but make sure that there is an entrance to get rid of the drones.

The bees in the top box will behave as though they are queenless and thus build emergency queen cells. For more information on this read Joe Graham's chapter "Management for Honey Production" in the 1992 edition of The Hive and the Honey Bee.

I have seen lots of beekeepers keep the queen in one box on nine combs. On a theoretical basis, this provides enough comb space for the queen. But the fact of the matter is that some of the cells are full of honey and pollen and are not available for the queen to lay in. If you are using two full depth brood boxes, you have full time manipulation moving individual combs about so that the queen has enough room. This manipulation is essentially moving full combs of honey and pollen down into the bottom box and to the outside, and moving old sealed brood combs to the second box and the combs with young brood down into the middle of the bottom box. This job has to be done every ten days to stop the colony from getting bound up and wanting to swarm.

Now let's look at the different methods for a bit:

1. Keep the queen in one box, and place an excluder on top before adding any supers for honey. The colony is almost guaranteed to swarm even when you cut queen cells every ten days.

2. Keep the queen in two full depth Langstroth boxes. Put an excluder on top of them and add honey supers. The problems and amount of time spent are eased a bit but there is much manipulation of individual combs. You will get a few swarms but, by and large, not very many.

Now a third method, which was developed by Dr Farrar in Wisconsin and adopted by Dr B Furgala in Minnesota, is to use three brood boxes. The manipulations before the honey flow starts and after it starts are a bit similar. As the spring build-up takes shape the bees will almost abandon the bottom box. To stop any swarm preparations, place the bottom box on top and the queen will run right up into it laying her eggs. She has lots of room.

Now I am going to assume that your bees are into the middle of the swarming season. Here in America it could be March (in South Carolina) or May (Wisconsin). During this period, if the weather is fine, the bees will gather enough honey to make a net gain most of the time. However, if the weather is not very good then the bees may need to be fed. The manipulation should be to reverse the top and bottom boxes in two weeks. That is, of course, assuming that the queen has filled that top box with eggs and young larvae. The bottom box should be filled with emerging sealed brood.

The manipulation is continued every ten to fourteen days

until you have to add supers. You will then find that you will not need to use a queen excluder because the queen will have ample space and will not feel crowded. As you know, a queen excluder is a rather expensive piece of equipment and it is easily damaged and gunged up with beeswax and/or propolis. However, as a precaution to prevent the queen from entering the supers, put two or three combs partially filled with honey in the bottom most super - that should stop her from going upstairs.

To come back to Dr Tew's talk. He spent a lot of time speaking about the excitement of a swarm; the catching of it an then putting it in to a hive. A long time ago I decided that swarms were not exciting and I refused to catch them unless they were very easy to collect. If they were up a tree or required me to use a ladder, forget it. No swarm is worth a broken leg or a bad fall.

Actually, I manage my bees so that they don't ever swarm. They receive new queens every year and frequently twice a year. I have raised queens for a long time and if and when I see a queen not performing up to par, off goes her head! If you are a beekeeper who doesn't normally raise queens then you should be doing so - and using breeding stock that is resistant to varroa and AFB. By doing this you will be able to get rid of most of the drugs you have been using. Colonies that have come on well using any of the above methods and have maximum populations before a flow starts should be able to spare a pound or two of bees to form nucs for any queens you raise.

The three storey hive has several other advantages. For winter stores you need more than 60 pounds of honey and this is pretty easy to achieve with a three unit hive. If the hive has 90 lbs , so much the better; just don't steal any of that honey. Also, when the three boxes are reversed frequently, the bees will usually accumulate more pollen which will be used in the spring build up.

Lastly, the queen will produce a large population of bees for over-wintering. After the autumn weather sets in and

SECTION 1. WHAT DO BEES WANT? 23

the colony begins to cluster there should be a population of 50,000 bees - this will drop though to 30,000 by the spring. A final note for successful wintering - put a one inch auger hole in the top brood box to give bees a flight hole, for during periods of heavy snow the bottom entrance can become completely blocked trapping the bees in the hive.

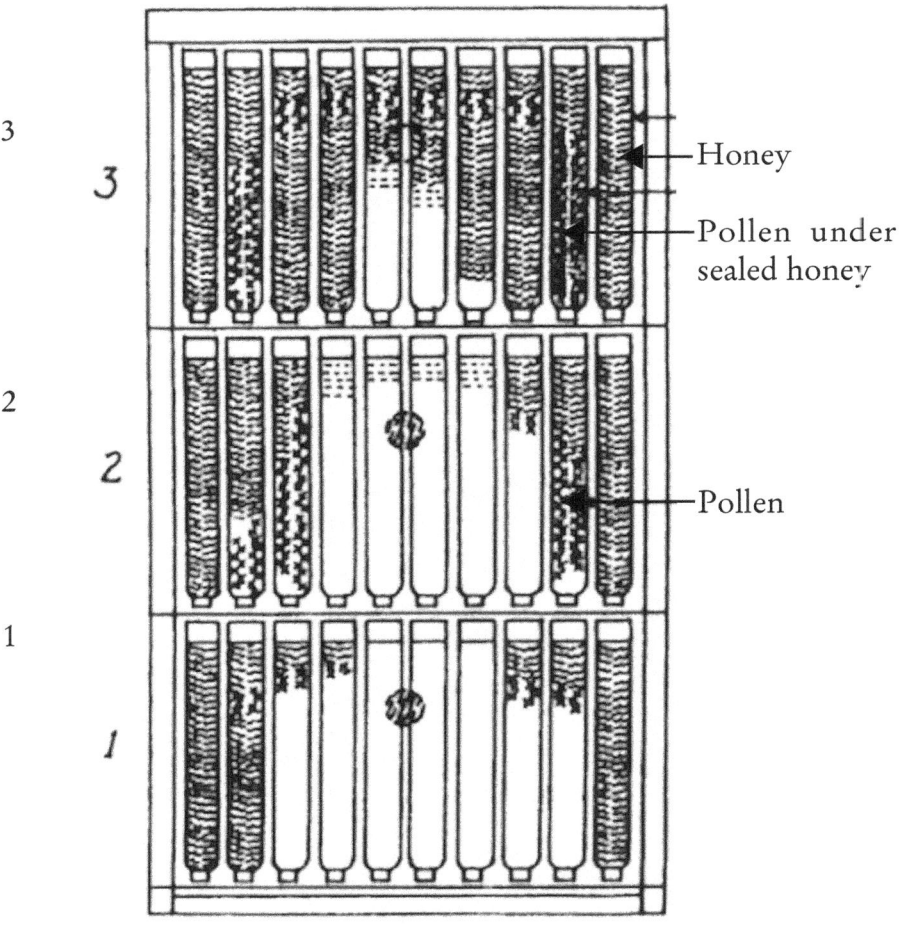

Wintering Productive Colonies

SECTION 2.
HEALTH AND PESTICIDES

BKQ 38
Summer 1994

Artificial Status of Bee Health

Recent editorials in Bee World by Matheson of the IBRA concerning World Bee Health illustrate a major problem facing beekeepers by various government officials and regulators. These problems are generally put into the bracket of non-acceptance of scientific information. The problem is exemplified particularly by the regulation or law concerned with "notifiable diseases". And I think this is very unfortunate.

I provide the following examples of current government attitudes which are incorrect and burdensome to beekeepers:

• Bee colonies with American Foulbrood (AFB) *Bacillus larvae*, should be destroyed by burning. Treating AFB colonies with drugs suppresses the problem, combs so treated will then be transferred into healthy colonies spreading the disease. And worse the disease will become resistant to the drugs in time and there is a danger of contaminating honey.

• Bee colonies infested with the mite *Varroa jacobsonii* have to be treated with a chemical or all colonies will die.

• Bee colonies infested with EFB, chalkbrood and other

diseases should be destroyed or the combs replaced regularly to reduce the incidence of the disease and so on.

All of these examples show what I call essentially a "burn and destroy" mentality.

Since the plague of AFB has been with us almost a hundred years since its identification by the American scientist, White, and it shows no sign of going away this article will be concerned primarily with that problem, that disease. Actually a great deal of research has been done on AFB employing many scientists world-wide for years. I shall not review this work here as it has already been reviewed in many books and pamphlets which are given wide distribution. What I will do is to present a few known facts:

• The disease is not specially infectious because feeding sugar syrup or honey to colonies which contain 500,000,000 spores rarely will cause an infection. In fact the only sure way to cause an infection of a colony of bees which is susceptible to AFB is to place a comb or sample of a comb with at least 79 scales of AFB in the colony.

• Primarily young, less than 24 hour old, larvae are infected and they require rather large doses of spores but genetic variation exists to the number of spores which cause the young larvae to succumb to the disease.

• No country or state has succeeded in eliminating the disease with even the most intense and expensive inspection programs, but some have come close. In the approximate 30 years up to 1960 several states in the USA, notably New York and Ohio which tried to inspect every colony of bees in the state twice a year reduced the incidence of disease to less than 1 % of the colonies inspected.

• Certification procedures requiring inspection of apiaries to be AFB free before shipment of combless bees and queens is a farce and makes no sense as it is impossible to transmit AFB in the candy of queen cages or in the syrup which accompanies caged package bees.

• Genetic resistance to AFB has been documented many times since it was first shown by Park in 1936. (Disease Resistance and American Foulbrood, American Bee Jour. Jan 1936 and Disease Resistance and American Foulbrood, results of second season of cooperative experiment. American Bee Jour. Jan 1937) The science of genetics has made great strides since 1936 and Rothenbuhler demonstrated the existence of two genes which determine hygienic behaviour (HYG) which is one of the mechanisms responsible for AFB resistance, and there are others.

• The presence of HYG in bees can be determined easily and cheaply by inserting a sample of freeze killed brood into the brood nest. Fast removal indicates HYG, i.e. 48 hours or less, 3 to 5 days needed to remove the dead brood indicate traces of HYG and bees taking more than 6 days are non-HYG. However this is a comparison test and in my experience no more than 10% of colonies have ever tested HYG.

• It should be the purpose and objective for all beekeepers to increase the gene frequency of genes responsible for HYG behaviour. With cooperation and much education and testing and replacement of non-HYG queens the gene frequency would be changed drastically in favour of HYG stock.

• Stock that is HYG would not only be resistant to AFB but also EFB and is the only way at present to eliminate the fungus, chalkbrood. Additionally of perhaps greater importance, bees which are HYG uncap and remove pupae infested with the varroa mite.

How can this be accomplished? First of all the relevant government agricultural sanitary officers have to be trained in the use of the test for HYG by inserting freeze killed brood into the brood nest. Regulations should be changed requiring all persons who raise queen bees for sale, to raise daughters that have tested positive to HYG. All beekeepers should be urged to test all their stocks every year for HYG and remove all those queens which are non-HYG. The issue is important and deserves immediate attention.

As a concluding part of this program after it has been introduced and implemented for at least 5 years, all colonies of bees should have inserted into their brood nest a comb sample containing at least 75 scales of AFB which will eliminate all traces of bees susceptible to AFB, EFB and chalkbrood and a partial control of varroa mites will have been accomplished. You think this is drastic? Compare this with the drastic reaction of the public toward supposedly "pure honey" when traces of a foreign chemical, terramycin or fluvalinate show up.

And last I am not talking completely theoretically as I have been managing my colonies this way for the past 15 years. I accept gifts from two of my fellow neighbour beekeepers of badly diseased (AFB) hives. I kill the queen and replace her with one of my own and feed terramycin until the combs are all cleaned of the disease. I then distribute these diseased contaminated combs to my other hives. As of today, July 1, 1994, I have seen no disease in any of my colonies or the gift hives received in well over a year.

BKQ 40
Winter 1994/5

The Origin of Pesticide Application by Plane

A Bit of American History as it Effects (Some) British Beekeepers

Back in the 20s and 30s the largest entomology laboratory in the world was established and maintained in the small village of Tullulah, Louisiana by the United States Department of Agriculture for the purpose of controlling noxious insects on cotton, Gossypium sp. I realise British people know little of cotton but it is a very interesting crop and is attacked by more different insects than any other crop of my knowledge. Cotton is in the same plant family as the garden hollyhocks, Malvacaeae, and the blossom is very similar except those I have seen on cotton are bright yellow.

At that time the only insecticides in use were the arsenicals, mostly calcium arsenate and used as a dust. Mules pulled wagons through the fields that were loaded with this toxic chemical and black laborers scooped it out and spread it on the cotton plants. This was done hoping to kill the cotton boll weevil along with many other noxious insects. It was extremely labor intensive and dangerous to the laboring crew. At that time the growing of cotton was a very labor intensive /crop and few farmers had cotton patches larger than 10 acres and to get any crop at all the farmers all had to dust their cotton in a similar fashion. Also at that time cotton was the principal cash crop grown in states in the South Eastern USA.

During the early 30s the director of the laboratory thought there should be a better way and wondered if an airplane could used to distribute insecticidal dust on the cotton fields. This would cost quite a bit of money, it was during the Great Depression, and so he appealed for funds from his superiors in Washington, DC, headquarters of the USDA. His request was turned down and he was informed that this was a very silly notion and that he should forget it.

In the village of Tullulah (directly across the Mississippi River from Vicksburg, Miss) were some young fellows with an airplane. The director of the laboratory went to see them about the problem anyway asking if they would be interested. They were. The director then misappropriated money from his laboratory which was used in modifying the airplane so it could be used to apply insecticidal dust.

As you know the misuse and/or misappropriation of

government funds anywhere in the world is a very serious offence. The officials in Washington heard about this work and ordered an investigation by the government authorities who were auditors and after some two years of investigations reported back to USDA headquarters that: yes, money had been misappropriated by the laboratory director. All of this money had been used to modify the airplane and the modification to the airplane had been successful and was now being used by farmers to apply insecticide to their cotton fields.

The director of the laboratory was fired for his foresight and imagination which led to the formation of the crop dusting industry that we know today. In addition, this little crop dusting outfit in Tullulah, LA, grew and grew and evolved into a great airline which today is known as "Delta Airlines". When I was employed at the USDA bee laboratory in Baton Rouge, LA, I had the good fortune to meet and visit with one of the entomologists who had been employed at that Tullulah laboratory and got this information all first hand. And during this time from 1950 until 1965 many of the members of the Delta Airline board of directors were residents of Tullulah, the headquarters of the airline before it was moved to New Orleans and then to Atlanta, GA., where it is today.

I suppose the moral of this story is that if you have ability and imagination and you are interested in entomology (bees) as a career you shouldn't work for the government. Well, this was all brought about by the article pages 89 and 90 of The Beekeepers Annual 1994, entitled "On the right track" by John Kinross, and his reference to Gossypium and the photo of the ancient airplane in the Hive and Honeybee.

SECTION 3.
QUEEN BEES AND BREEDING

BKQ 53
Spring 1998

Queen Rearing without Grafting

There have recently been introduced into the many and varied things beekeepers spend their money on, several new techniques designed to collect eggs from the queen in a device which can then be used to rear a queen bee. And when I use the word "recently" and "new', that means in the last twenty years. Don't get me wrong, I have nothing against any of these things, I am sure they work as advertised, but the very best way of raising queens without grafting was designed by Dr Miller of Illinois almost a hundred years ago.

The beauty of the Miller Method is that if you already have as many as five hives of bees you already have all the equipment to raise queens using his method. You don't have to build, make, or buy anything, you just let your bees do all the work. All you have to do is know how to count, you know, 1, 2, 3, and so on up to 10 or 20. And you will have to sacrifice your honey crop from 2 or 3 colonies, or most of it.

I have organised this in a step by step procedure, so you will have no problem figuring out what to do next!

1. This is to select your breeder queen, your queen mother. Don't make a big deal out of this, but I do suggest you test your queens and select the one with the best record of hygienic behaviour.

2. Wait until you have nice spring weather and you see many flying drones and when you are working your bees you see much sealed and unsealed drone brood. This Miller method works all the time, but if you try it during an intense honey flow you will have to modify it a bit. (Intense honey flow is defined as a gain in weight in excess of 10 pounds a day over several days). Just off-hand you would think that during a honey flow would be the best time to raise queens, but this is not correct. When the bees are in a honey flow they go crazy and don't want to do anything else except collect nectar and turn it into honey. You can stop this by moving the queen rearing colony about 5 - 10 feet to lose the field force, placing a dummy hive with a frame of brood to collect the bees who are so desirous of collecting nectar.

3. You now want to adjust the size of the colony that contains your breeder mother queen. Ideally, this unit should fit very well into a 10 frame box with lots of room for the bees left over; don't crowd the bees or queen. And you don't want a bunch of old field bees here with the queen; you want young bees which have just recently emerged. If there is no honey flow on you will need to make some arrangements for feeding bees as you want them in comb-construction mode.

4. Next, you prepare an empty frame with some pieces of foundation cut into sort of triangular shapes and insert this frame into the centre of the brood area. If everything has been done correctly the bees will immediately begin drawing out the foundation and adding comb to the bottom and sides of the foundation pieces. However, from my past experience with bees, they rarely do exactly what you tell them to, when you tell them to do something. So you should examine this new frame with the bits of wax foundation within a couple of days to see what is happening.

If all is perfect you should see that the strips of foundation now contain some partially-built comb and some eggs, and at the top, some newly stored honey or sugar syrup. If this is not the case you may not have enough young bees in this hive which means that you have to add some, or that they need to be fed with additional sugar syrup.

Continue to examine this frame every two or three days until you see that the queen has laid eggs in the area that was foundation and that the bees have started elongating the original strips into new comb and that the queen is beginning to lay eggs there also.

You have to be prepared to expect that the bees can do all of this in less than 24 hours, but it will take at least a week before the bees could completely fill the frame with new comb - so at this point there is no hurry to do anything.

Ideally, over a period of several days, the bees are adding a bit of comb every day and the queen is placing eggs in the comb as the building progresses. If you remove this frame for an examination (do not make this examination in full sunlight as sunlight is very harmful to eggs) on the 5th or 6th day you should see a gradation in the new comb, from top to bottom, of young larvae, hatching eggs, and newly-laid eggs.

5. Take this comb, brush off all the bees and lay it on a flat surface. With a sharp knife, cut the the lower portion of the comb off containing eggs and discard. Now go back to the hive and find the queen and cage her with attendants or make up a special nuc for her. At any rate make the colony queenless. Then insert the new frame with very young larvae right where it was in the brood nest.

The bees prefer to build queen cells on new comb rather than old comb so that all the queen cells (almost; with bees you never say "never" or "always") in this now queenless colony will be built on new comb. At this point you have to know how to count to ten (10). When you get to ten days from the date of the original manipulation of removing the breeder queen you have to bring the comb with cells in and remove them and place each in a queenless nuc to allow the virgin to emerge and to mate. Again, mark the date on your calendar when the cells go in. Day one or two, expect the virgin to emerge, days five, six, or seven, weather permitting, the virgin may fly and mate and, if all is perfect the new queen will be laying eggs on the 14th day.

Of course, after the cells have been removed the breeder queen can be returned to her original colony and it can be returned to its original position.

There is very little that can go wrong with this queen rearing method, but there are two or three things you need to watch out for:

(i) The breeder queen will not deposit her eggs in the new comb during a heavy honey flow: get rid of the nectar collectors.

(ii) The bees build drone cells and the queen lays drone eggs: tear them out, tell the bees to do it properly, and give them some drone comb to play with (they obviously want some).

(iii) No honey coming in and you forgot to feed them so there is no, or little comb that has been built: feed them.

There are countless ways of making up nucs for queen cells. I suggest you see pages 1012 -1017 of "The Hive and the Honey Bee", edited by Joe Graham (1992), or my book, "Breeding Super Bees", page 10.

BKQ 86
November 2006

The Right Bees for Queen Rearing
Letter Re 'Queen Rearing Issue' BKQ No 84 July 2006

Dear John,

Mr Dawson's article 'Queen rearing Made Easy' in BKQ No 84 makes the usual error in wanting strong bee colonies to rear queen bees. Actually what you want and need are bees of the right age. Dr Peng who obtained her Ph D concerning queen rearing in Canada showed that you need about 250 bees of between 8 - 10 days old for each queen cell raised. If you have a normal colony with a normal queen she should be laying 1,200 eggs per day, consequently you would have 1,200 bees in the hive that are eight days old, 1,200 at nine days old, and 1,200 bees which are ten days old. These bees are the ones that will raise good queens. Flying and foraging bees do nothing to enhance making better queens. In fact, during a honey flow I frequently move the queen rearing hive to about eight feet away to lose the field bees so that the queen feeder bees are not disturbed by all that activity.

Pollen and open honey need to be in close proximity to the queen cells. Open honey does not mean that you scratch the capping from sealed cells because this causes the bees unnecessary work cleaning up the mess. If pollen is scarce make up a cake using pure bee-collected pollen, sugar and a bit of brewer's yeast. Place the cake directly over the frame with the queen cells. Queens: poor mating and laying (re Letter, Roger Patterson).

There are many causes:

Firstly, the practice of placing miticides in the hive to control varroa. Research at Baton Rouge USA Bee Laboratory has showed this to have a very detrimental effect on queen quality as well as drone quality.

Secondly, commercial queen breeders frequently have more queens available than orders for them so they put the surplus queens in a bank. Work in Poland by researchers at the Warsaw bee laboratory showed that frequently the bees damage the queens held in these queen banks. In particular, the queens could lose their foot pads which enable them to walk on glass. Many such queens are frequently superseded by the bees.

Lastly, queen mating nucs and queen rearing hives are usually heavily infested with the nosema parasite. Research by Dr C L Farrar about fifty years ago, as well as others since, have shown that the drug Fumidil B fed to queen rearing colonies and to nucs, prevented the queen from being infected with that disease.

Steve Taber is most likely right (he usually is) in that it doesn't need a strong colony to rear queens, but I was trying to offer to readers an EASY method of queen rearing, not necessarily a scientific one, and it is EASY to do in a strong colony as I described. David Dawson. Manitoba, Canada.

BKQ 49
Spring 1997

Multiple Mating of Honeybee Queens

Every now and then I get asked the question by some beekeeper, "How do you know a queen mates with 5, 10 or 15 drones when you can't watch what is going on?" Good question and that is what this article is about. Actually, this was my very first problem I undertook when I was first hired by the USDA way back in 1950. But first let's look a bit at what was known at the time. Two hundred years ago the Swiss naturalist, F. Huber, discovered the virgin queen bee mated and that after she mated she returned to the hive with a 'mating sign', a portion of the sexual organs of the drone protruding from her abdomen. Huber reported seeing at least one queen returning from mating flights that had 2 mating signs, indicating she had mated with at least two drones. This last observation was largely ignored and when you look at the beekeeping literature from 1840-1940 there is nary a mention of the possibility of a queen's multiple mating, in fact you will find that many bee scientists, primarily German, were proving that it was impossible for a queen to mate more than once.

Then in 1944, W.C. Roberts showed that 50% of his virgin queens under observation had mated with at least two drones and in the Soviet Union, V.V. Triasko in 1951, showed that virgin queens which returned from a mating flight had three to four times the volume of sperm in her oviducts as there was in one drone.

That's the way it looked when I was first hired in 1950 to be Dr. Mackensen's assistant, to raise his queens and drones and to take care of his colonies and nucs so he could do his research. Actually this left me with a lot of time on my hands and with lots of bees and other available equipment. In addition there were two events or items I was interested in; one was to learn the technique of artificial insemination (AI) and frankly I was having a terrible time. In those days the equipment was pretty crude and I am a very slow learner. The second item was my fascination with the bees' body colour, a mutant called cordovan (cd), I thought it was just the most beautiful bee I had ever seen and I wanted more of them. During this long period of several years of taking care of Mack's bees, about 50 colonies, I had requeened a lot of them with cd queens I had inseminated or cd queens that had naturally mated. All of a sudden I began to realise that almost all the cd naturally mated queens had both wild type (in most genetic terminology the usual form is called 'wild type' as opposed to a mutant form in this case cordovan) and cd worker progeny. Back at my office I began to diagram possible mating events.

Here is what we knew about the cd mutant; that it was recessive to the wild type (wt), and that a mating of a cd virgin to wt drones would produce worker bees all wt in appearance. A mating of a wt virgin with cd drones would produce all wt (in appearance) worker bees. Now then, if a cd virgin mated with a cd drone and also a wt drone, both wt and cd worker bees would be produced.

While studying at the University I had taken many courses in mathematics, probability and statistics. I didn't do very well in those courses but I really did learn some things; now I had to put my school knowledge to use in solving a mathematical puzzle. But while I worked at this puzzle I had to set up the apiary to get some actual numbers to work with.

The first step was to requeen the apiary with naturally mated cd queens who would produce, each queen, 100% cd drones. I did this without consulting anyone, I knew Mack wouldn't care as all he wanted were bees to stock all his nucs in the spring. Two months later I began grafting and raising cd virgins that I mated down at that apiary. All I had to do then was to see what type of offspring the queens produced which I recorded, either all wt or all cd or a mixture of both. Altogether I raised 184 queens in this experiment where I recorded the offspring worker type. The numbers I had to work with were: 171 queens with both cd and wt progeny, 13 queens with just wt workers and there were no queens which produced cd progeny only. Now I had all winter in my office to try and figure out this puzzle.

I don't really think readers would be interested in all the mathematics involved in solving this puzzle but if you are consult the references at the end of the article. However you can see that the more matings a queen has the less chance she has to always mate with the same genetic type and in my math I came up with a total of 7.5 matings on average for each queen.

This work generated a lot of interest in several quarters so much so that the work was repeated in Illinois by Dr. Bud Cale and in Canada by a graduate student of Dr. Farrar's, Don Peer. Here is their data from their mating studies.

As you can see these people made a greater effort to control drone populations than I did but their results were essentially the same in that the data indicates a number of matings, not

PROGENY OF QUEENS				
PERSON	PLACE	PROGENY BOTH FORMS	ALL CD	ALL WT
Cale	Illinois	241	6	3
Peer	Algon Park	162	1	1
"	Ottawa	293	1	9

just one, two or three.

After publishing this data I continued to think about the problem and discovered a small error in my math. So I worked and worked on this new problem and decided to take it over to the math department at the Louisiana State University where the USDA bee lab was then located. Then I ran into another problem.

I met a young scientist who had done a lot of work in theoretical statistics which was the kind of person I wanted to talk to but when I began talking about honeybee queens mating and mutants and wild types his eyes just glazed over so I had to, on a spur of the moment, change the entire way of presenting the problem, here is how it came out. Given; a large container with black and white marbles, (the marbles represent the cd and the wt flying drones) which are shaken up, you reach in with a scoop and withdraw some and record whether the marbles are all black, all white, or of both colors. At the end of several hundred of these samples the question is asked "What is the average number of marbles in each scoop sample?" Of course, 'scoop' represents the queen on a mating flight. Then, my math friend, James Wendel, woke up and showed intense interest. We wrote a paper on this brain work then and with it came to the conclusion our American queens were mating about 10 times.

For those of you who have access to a library and would like to look at the original references, here they are:

- Taber, S. 1954. J. Econ. Ent. 47:995-998
- Taber, S. 1955 Amer. Bee J. pp 474, Dec.
- Taber, S. & J. Wendel. 1958 J. Econ, Ent. 51:786-789

BKQ 75
November 2003

Finding Queens

This article will tell you how I find queens in different situations because the different situations require different techniques but first remember that:

1. Queens run away from the light unless they are very old;

2. Queens are most likely to be found on combs where you see lots of eggs;

3. Always use the barest amount of smoke (instead of a smoker you could use a mister which contains mostly water, but a little sugar added will be of benefit) .

Finding Queens in Nucleus Hives

The first situation in which one might be looking for a queen is in a nuc containing from one to three combs of brood and space for two or three more. Have the sun on your back or directly overhead. Remove both outer and inner covers. A cloth, plastic or burlap is better than a piece of wood and masonite for an inner cover. Before you remove any comb, look at the side of the comb facing you, then slide it over to the empty space immediately looking at the other side of the first comb, then the side facing you of the second comb. In most cases, you will have already found the queen without having to look at the third comb. And, you should notice that the queen runs to find the edge of the comb and will crawl underneath if you give her a chance.

Of course, it doesn't always work that way. I had a very embarrassing moment once when I worked at the Tucson, Arizona, USDA Bee Laboratory. My bee yard was on the

University of Arizona farm about a mile away from the laboratory. I used that place to raise all my queens, to do all my artificial inseminations, and many other experiments. I frequently had visitors there also. One day a couple of carloads of beekeepers arrived to see what I was doing. As they arrived, I was checking on some queens I had inseminated to see if they had started laying eggs. All these guys walked in on me wanting to know what I was doing and I replied that I was looking for the queen. I had all the nucs on posts about waist high so that I could work better with them. The nuc only had three combs in it and I looked them all over several times, then took them all out in order to examine the bottom and side walls. No queen. Then, one of the beekeepers said, "Is this what you are looking for?" He held a queen in his hand, very dead, as I had stepped on her. That was that.

When you go through a nuc quickly using very little smoke and still don't find the queen, you have to assume that the queen is running away from you. Go through the combs slowly the second time and hold the combs over the box, so if the queen falls she will fall into the box. Another thing which can cause a bit of trouble is that the bees will hide the queen by a bunch of them covering her.

When this example happened I was operating a commercial queen-rearing establishment in Vacaville, California, and had about a thousand nucs being used to mate virgin queens. I employed a number of people to help me, both boys and girls; they all had some experience of working with bees except one girl who was the twin sister of Gail, who had worked with me for about five years. When Gail called me and asked if she could bring her sister, Nancy, all I asked her was whether her personality was similar to her own. When she said yes, I immediately said to bring her.

When you are raising queens commercially, you can't be concerned with the weather - whether it is raining, snowing or even if it is 115 degrees F (46C). The work has to be done NOW. So, when you are in the rain all day, with good rain gear on, you still get all wet. At such a time you do not want to hear any complaining. These two girls were always happy and cheerful, regardless of the weather.

One day I walked into the nuc yard to observe what was going on and I came up to Nancy. She was hunting for a queen to cage so that it could be shipped out. The bees were hiding the queen by covering her up and I saw this brand new apprentice beekeeper take her finger, not covered with a bee glove, and push the bees away very gently so that she could pick up the queen. I asked her "Who showed you how to do that?" She said, "Nobody, it just seemed the obvious thing to do."

Colony With Just One Box for Brood

The second situation in finding a queen is when you have a small colony with all the brood in one box. The procedure you follow is about the same as finding a queen in a nuc box. Get comfortable with your back to the sun and take out the second comb and examine it and place it on the ground near your feet. Take the comb out which is next to the box wall next, examine it and place it next to the first one. Use just a minimal amount of smoke as too much will cause the bees and queens to run. You should be down on the ground so that you can move your hands across the combs horizontally, not coming down on them from high. Move slowly and deliberately and remove the third comb. While you are doing that, look at the side of the fourth comb facing you. Then examine the comb in your hand for the queen. Hold the comb over the hive so that the queen, if she falls off, will fall into the box. Continue going through the hive until you have examined all the combs.

If you haven't found the queen, do it again and this time examine the bottom board and the sides of the hive. If the bees have begun to run, take all the combs out and look under the bottom board and that is where you should be able to pick her up. It also tells you that you used too much smoke - you have to be able to take a few stings.

Colony With More Than One Box For Brood

The third situation is a hive with brood in multiple bodies. Take the hive apart, placing each brood-containing box over a queen excluder and go through each box as described in situation two. You will most likely find the queen on the comb that has just been filled with eggs.

Finding the Queen in a Bad-Tempered Colony

Now for the last situation which confronts every hobby beekeeper once in a long while. You have a feisty colony that is hard to work because the bees suffer temper tantrums as soon as you open the hive. Get all suited up and have on your gloves, have the smoker going full speed ready to to blast to the moon. Take the hive apart and then put it all back together again, except put a queen excluder between each box. Now pack it all up and go home and wait for at least five days. Try to get some other person to give you some help. That is one of the purposes of your bee club, so that you can learn and help each other. So, after five or more days, go back to the colony that you had actually thought of torching and slowly take it apart. Look in each brood box until you see young brood and eggs. Now you know where the queen is and is not. Quickly go through the nine or ten combs as described before. When you find the queen grab her and remove her head, throw her carcass out in the grass and introduce a more docile queen to head that colony.

Remember again what I have said: when hunting for a queen use bare hands and very little smoke. If possible work your hands horizontally across the hive and not vertically. Convince the bees and yourself that you know exactly what you are doing and that you are in charge. Exude confidence. Remember this hobby is fun. One last word: You should be working your bees without gloves most of the time. When you are taking off honey or doing something violent to your

SECTION 3. QUEEN BEES & BREEDING 43

1

2

3

1. This shows the correct position for the beekeeper to adopt when looking for queens. Either squat, kneel, or sit so you are in a comfortable position to move bare hands across the top bars and you can see if the queen is on the adjoining comb.

2. The incorrect way to look for queens. The back is bent and is under strain and the beekeeper cannot see the bees on the adjoining comb.

3. Do not use gloves when just looking for queens. Here the beginner, whilst adopting the best pose, is having some difficulty partly because the gloves are too big for her. The ends of the gloves get trapped when replacing the frames and lack of sensitivity at her fingertips leads to clumsy movements which annoys the bees.

bees like moving them or shaking them into packages, wear gloves. However, when you are specifically looking for queens do not wear gloves.

BKQ 37
Spring 1994

Breeding Bee in Villebrumier

I thought some of you beekeepers would like to hear about my program and regime of work since it is a bit different than that of anybody else. I do not produce more honey than I can eat as it is a lot of work producing honey and it is no challenge to me as it is so easy to do. A lot of beekeepers keep bees because of the different challenges they present, as I do and not for the potential honey they could produce. My challenge is to breed the bees I want to have in my hives that will do there what I want them to do. I want bees that are gentle and nice to work with and bees that have the hygienic behaviour (HYG) characteristics and also bees that do not have that characteristic. So here is my program of work for this summer.

It is now mid-Feb., winter is almost over, buds on the wild prunes are swelling, soon the bees will be out gathering pollen and raising brood. In a few weeks I will begin testing all my colonies and overwintered nucs for HYG by inserting a small piece of freeze-killed sealed brood into the centre of the brood nest and recording how many days it takes the bees to uncap and remove the dead brood. While I am doing that I will carefully observe and record the temperament of the bees in the colonies I am working.

From this exercise I will select 2 or 3 queens whose bees are HYG and again 2 or 3 queens whose bees are non-HYG. Since I have been already doing this testing and selection for several years I know that I will have no difficulty in picking up bees that have both these characteristics.

These queens will now serve me as the breeders for the year provided they passed the test of being reasonably nice to work with.

Before going further I need to explain why I want these two types of bees. Daughter queens raised from queens that are HYG will be used to requeen almost all my colonies. Some of these queens will be inseminated artificially but most will be mated naturally. All daughter queens raised from mothers that are non-HYG will be artificially inseminated with drones from queen mothers that are also non-HYG. These queens will be used in an experiment that I have in cooperation with a microbiologist at the Tucson USDA bee lab, Dr. Martha Gilliam, and I will explain more of this experiment later.

After queens have been reared and mated, some from HYG mothers will be used in colonies to replace those queens I don't like because of their temperament or because of any other reason. I will inseminate about 40 queens or more, some will be HYG x HYG but most will be non-HYG x non-HYG. The nucs that contain these queens will be relatively isolated but I know there will be drifting bees from one nuc to another. However, I know that most of the house cleaning in a bee hive is done with the younger bees, not by the flying bees, so a few drifters won't hurt anything. These inseminations will be complete and queens laying by the first of June.

After this I have a period of rest waiting for the old bees to die and the inseminated queen's bees to take over all duties in the small nuc hives. The rest period for me will last about 6 weeks, or until about the middle of July, and during that time I will make some increase as I want a few more colonies and will see to it that the colonies I do have are managed and manipulated so they will store honey and not swarm.

All my colonies are in frames called 3/4 depth, or Modified Dadant, making manipulation relatively easy as the full boxes of honey will only weigh about 50 pounds. From about mid-May until the middle of August I usually have a continuous honey flow from one plant species right after another and as the big colonies fill a super with honey I place it on the bottom board to get rid of it.

Why not extract it you say? To start with I don't have an extractor and I do have lots of nucs that get hungry at a drop of a hat. If you have never raised and mated queens in a bunch of nucs you have no idea how much honey or sugar the nucs need to survive. Many times they even have to be fed during a honey flow. So the honey the big colonies make goes on the bottom board to save it to feed the nucs at some later date.

As the bees don't like honey below the brood nest, they will proceed to slowly move it back up again, no problem, for as another super is filled I place it on the bottom and the first one placed there, which will now be partially full, will either go on top if it is more than half empty, or remain down below if more than half full. You can see that the supers of honey below the brood nest act as a feeder between the honey flows to the colony so they will raise good queens for me and drones all summer long. Secondly, putting the full honey supers down like that moves the brood nest up higher making the regular 10 day inspection much easier on my poor back.

A third advantage to me is when I go into a hive to examine, inspect and manipulate combs, I no longer have to lift off a bunch of supers. This manipulation also puts the majority of bees that want to be irritable down below in the honey supers.

I have been managing queen rearing colonies like that, whenever they would start making any surplus honey, for about 40 years. But now I also use the system to store honey on the colonies not used for queen rearing to make the colonies easier for me to work.

Now that the six weeks are passed and it is mid-July I again test all the new queens, all those naturally mated and inseminated, with frozen samples of sealed brood. Just because queens have been raised from queen mothers who have been tested as HYG or non-HYG doesn't mean that their daughters will have that characteristic. The genetics of HYG have not been worked on except a little bit by Dr. Rothenbuhler and he looked at very special cases which were the result of a lot of inbreeding.

Hopefully, at the conclusion of these dead brood tests I will have about 20 queens that are all sisters and mated to brother drones that were supposed to be non-HYG x non-HYG and are, as the test shows.

After the tests are all complete I make up a special pollen patty mix of equal parts of pollen and sugar, about 5 pounds of each, then add a little water until it all sticks together nicely. I next take 10 black chalkbrood mummies and 10 white ones and grind them up until they are very finely ground and add them to the pollen mixture, mixing everything together very carefully, then I add just a little brewers yeast to the mixture to keep it from running. About a quarter of a pound of this mixture is placed on the top bars of each of the non-HYG nucs right above the brood.

Five days later I examine each of the 20 nucs and examine the brood, checking for the presence of chalk brood. Almost all of the nucs will have from 20 - 50 cells of chalkbrood in one frame, but strangely, usually about 1 in 10 of the nucs will not have any chalkbrood mummies at all. Why are there no chalkbrood mummies in these nucs which have all been tested as non-HYG? What makes them different?

Dr. Gilliam and I have been working on that problem for about 10 years. She has found from some of the many hundred samples I sent her, several bacteria and moulds present in those nucs which inhibit the chalkbrood fungus from developing and growing. She has isolated these organisms, grown them in the laboratory in Tucson, and now sends me cultures of two of them to test in the 18 nucs which came down with the chalkbrood fungus.

About mid-August I will receive this material from her and again I will mix up the pollen-sugar-Chalkbrood mummy mixture, but this time I add to one sample a material received from Dr. Gilliam, and then a second batch receives a second material from Dr. Gilliam and the third and last batch receives nothing and is the control. There will be six queens tested with each of the 3 batches. Five days after placing the pollen mixture on the nucs and again at 10 days and 15 days, all chalkbrood mummies will be counted.

What do you think the results will be? I don't know. That is what science is all about, probing into the mysterious unknown.

BKQ 74
July 2003

Why Keep the British Bee? Why Indeed!

A letter in response to Ashleigh Milner's article in BKQ 71, November 2002, "Why keep the British Bee? An introduction to the understanding of honeybees, their origins, evolution and diversity."

Instead of all the subjective arguments presented by Mr Ashleigh Milner why he wants the British bee in his hives, why doesn't someone do some controlled experiments to see if there is any difference in any of the bees you have access to? Because to me sitting over here on the other side of the Atlantic Ocean, it sounds like pure, unadulterated prejudice.

I started keeping bees way back in 1939 and except for two years in the US Navy during the war, I have been in beekeeping ever since - meeting and talking to beekeepers not only here in the USA, but also in Europe, Africa and Canada. It was very apparent to me that most beekeepers were prejudiced about their bee race or colour of their bees years ago and with no real reason for it.

I studied bees at the University of Wisconsin in Madison under Dr C L Farrar. He was one of the very few people who did any stock testing to try to determine differences.

As his student I assisted in these tests. He measured square inches of sealed brood every ten days of the season beginning in April and ending in late July. He measured total honey production and winter consumption of honey by weighing every colony and every super that went on each colony and weighing it again when it was taken off.

Wisconsin winters are much harder than those you have there in Britain, for here the temperatures in January and February will rarely come above freezing and at night frequently they fall well below zero degrees Fahrenheit (-18C) and every winter would go down to minus 20F or lower. Italian bees did very well. In fact, the greatest difference was always determined not by the difference between races or stocks, but by the variability of queens from queen breeders selling the same race of bees.

What is needed there with you nice British folks is to do some very intensive stock testing. Weigh everything that goes on each hive and everything that comes off. Count the square inches of brood at least every two weeks. Weigh the hives when you put them to bed in the fall and again when they come through winter to spring.

Breeding bees resistant to *Acarapis woodii* is an easy thing to do with the use of AI. I did this while breeding bees during my stay in California but, again, this also needs testing. Without testing you will be standing still with your British Bee. When testing bees you need to have comparisons between races - testing one Black Bee against another Black Bee will get you nowhere.

BKQ 47
Autumn 1996

Brother Adam and the Buckfast Bee

Have I ever worked with the Buckfast Bee (BB)? No. Why? Well there are multiple reasons actually. Mostly I develop my own bee which I am interested in and I am not at all into the production of honey any more. I started my bee work working for large honey producers (large then) in 1941/2 who had thousands of colonies and produced tons of fine honey. During the war I flew aeroplanes for the US Navy and when that was over attended University to learn a bit more about bees.

There an awful shock hit me, I discovered that somebody else knew more about bees than me. Have you ever been 20 years of age and thought you knew everything you needed to know and of a sudden had to realise that some other person actually knew more? It is quite a shock. I worked with Dr. C.R. Farrar for my 4 year university degree at Madison, Wisconsin where he was interested in two things:

The first was the production of a maximum honey crop from each and every colony at the University, about 350.

Secondly he wished to evaluate which stock, race, breed of bee would produce the biggest crops.

Anyway by 1965 or so I had decided that the production of honey was really hard work, very messy work, and not very rewarding, so I concentrated on the study of all the bee literature, including that of Brother Adam, and my scientific bee endeavours. Now then, back to Bro. Adam.

The first thing to realise is that he single-handedly brought honour and recognition to beekeeping around the world and to the non-beekeepers amongst us. And that is important as we beekeepers have in the public's eye a rather unsavoury reputation as peculiar people. Secondly, he led the ranks of beekeeper naturalists who throughout history have made remarkable contributions to the study of the natural history of bees and this includes such as Dzierzon of Poland and Langstroth of America.

What exactly was the BB? Start out by reading his book 'In search of the best strain of bees'. In this trip, actually a number of trips, there he visited beekeepers and their apiaries in Northern Africa and in countries surrounding the Mediterranean, you will see he brought back to Buckfast

Abbey queens from many different areas and defined by Dr. F. Ruttner as subspecies of the bee *Apis mellifera*. Some of these bees he included in his breeding work in the development of his 'perfect bee'.

Ten years ago I wrote a book entitled 'Breeding Super Bees'. Does this BB qualify as a super bee according to my definition? Well all I can say is, maybe. My definition is and was of a super bee one that had been shown by suitable testing and breeding to be different from other bees available. This is where I have always had difficulty with Brother Adam. I first heard him lecture to a beekeeper group in Columbus, Ohio about 1963 and I have heard him several times since and I have read some of his writing. He talks and writes about 'testing' this cross and that one and that these were included and these were thrown out and he keeps doing this for many seasons and generations. So what is my problem? He never ever says or writes how these tests were conducted - amount of honey produced; resistance to the internal mite Acarapis? or temper or anything. As someone who likes to think of himself as a bee scientist I want to see numbers, data, how many colonies in each test and how much variation was there between tests and within tests. These things or items of data were never forthcoming from Brother Adam.

Another major problem I have with the BB bees is the extreme prejudice beekeepers have for and against some bees and most of the time it is, like all prejudice, totally unreasonable. For instance, they want yellow bees or black ones, or Carniolian, or dark Italian or in many cases those from a particular queen breeder.

Now in actual fact, worldwide tests comparing the productive ability of any races or strains are rare. A few years ago, the University of Minnesota ran a series of tests under the guidance of Dr. Furgala comparing queens received from different Southern USA breeders and found a slight difference in them. Then they also tested the Starline Hybrid against itself by testing the same, or supposedly the same stock but reared by a number of the different authorised Starline breeders and again they got a statistical difference between the different breeders. Again as many of you know a 'statistical difference' can be very slight, say between averages of 90 pounds for one group to 85 pounds for another and a 5 pound difference in honey production is not enough for any person to notice when removing full supers of honey.

I do not know how many bee colonies Brother Adam kept at the Abbey and managed for honey production but I would suppose no more than 350 colonies. As a student at the University when I studied under Dr. Farrar, he had 350 colonies all involved in stock testing. During the late spring and summer in Wisconsin this physical work associated with these few bee colonies, weighing all equipment on and off each colony, counting bee populations, counting brood production checking every 10 days on the marked queens presence, required the full time efforts of Dr. Farrar and between 4 and 7 students. Production studies were made in addition, on 2 pound packages of bees with queens received on or about April 15th, over wintered colonies and two queen colony units. Some of the stock under test was bought from individual queen breeders, others were developed using artificial insemination to produce certain hybrids and this work was done by Drs. O. Mackensen and W.E. Roberts. So what this paragraph is saying is that Brother Adam did not have this much help from young muscles and high technology using the most current genetic information available or the general expertise that comes from having a large University backing as well as the US Dept. of Agriculture. But, is it possible for one man to do it all alone? Yes, I think it is. But I think also that it is impossible for anybody to select bees and breed for increased honey production. All the issues which are involved in honey production are way to complex.

As an example, take dairy cows; they produce milk, cream and meat. Have you ever considered how many variables there are to milk production? And all these items, or a lot of them, have now been measured, like what they forage on what they are fed, how many times a day they are milked, various disease control measures and on and on and on. In fact if you look into it you will find an extremely unusual fact claimed by the dairy herd improvement associations everywhere in the world now, of how much milk production is increased in the next generation by using a particular bull's semen.

One time I had a conversation with Brother Adam about breeding bees resistant to the internal parasitic mite, *Acarapis woodii* because another beekeeper knew I had done this and brought up the subject. He wanted to know how long, how many years it took me, to which I replied 'three', and he said then that he had worked on that problem for 30 years or so. This characteristic is not difficult to breed for as has been shown by others, but you have to devise an adequate test, you have to compare the infestation levels from different stocks always with a starting point where all test units are equally infested. Again, all I could get out of him was that he had 'tested' different stocks and discarded those he didn't like. I was still left in the dark as to what exactly his 'test' was.

Anyway all of this is what I call 'nitpicking' as it is obvious to me and to many others that Brother Adam was a top flight beekeeper and tireless promoter of better bees and better beekeeping. And as my last word on the subject, I think our beekeeping world was much the better for his being here with us for his 98 years.

SECTION 4.
HONEYBEE BIOLOGY, BEHAVIOUR AND EXPERIMENTS

BKQ 44
Winter 1995/6

Bee Eggs & Larvae

It is time to think a bit about bee eggs, what we know and what we don't know and about bee larvae, too. And unfortunately, I can tell you right now that most of the things you have heard about eggs and larvae are just not true. So what I am challenging you with right now is information which you should check out with your own observation to see whether I am right or wrong. And I will tell you I have been wrong many times but every so often I turn up right.

OK, lets begin. Almost all books say be careful to not chill the brood as you will kill the eggs and larvae when working the bees in cool spring or fall weather. Don't open up the brood nest and expose the brood to the weather and so on. Well that is all a bunch of nonsense, so try this: take a comb with eggs and larvae, young ones, just hatched eggs, and just wrap them very carefully in a wet towel, not dripping wet, put it in the refrigerator overnight and put back in the bee hive with nice warm bees and nice warm brood about 24 hours later, marking the comb. Almost all dead larvae and eggs will be eaten by the bees in 2 days. So check the comb 3 or 4 days after that treatment.

Now some people keep their refrigerator too cold and there will be spots in it where frost will form. You can't let the eggs and larvae get that cold as it will kill them. But I will tell you I have kept bee eggs and day-old larvae on ice for several days then put them back in the bee hive and they were nice and alive. In fact I even raised queens from such larvae that was supposed to be all dead.

Over the years I have taken many thousand bee eggs away from bees and done some pretty weird things with them, like putting them in ice water, yes the eggs were dropped one at a time into the water where they float. Then to make them sink, I took an eye dropper and dropped water drops on them until they sank to the bottom of the vial. There the eggs stayed for a day or so at 0.5°C until I would dump them out onto a piece of damp filter paper in a petri dish where they then went into an incubator to see whether they would hatch, and many would.

What I found out, and what many other bee biologists had also found out, was that bee eggs will dry up and die unless they are kept in an environment of nearly 95% humidity. So one of the things you have to be careful of when handling brood combs is not to let the eggs and larvae dry out which can happen quickly. And refrigerators are extremely dry places so when you place a comb of eggs and larvae in there it has to be not only wrapped tightly with a wet towel but when I do it I tie it so no air can get into the brood to dry eggs and larvae.

The second way you can kill a lot of brood is to expose it to direct sunlight. Again, don't trust what I say, test it yourself. Take a comb with eggs and larvae and expose it for two minutes to direct sunlight, mark it carefully and examine it in 4 days time to see what you will find. For 15 years I had the privilege of working with one of the finest bee scientists in the world, Dr. Otto Mackensen. That period was from 1950 until 1965 and during most of that time Dr. Mackensen was working on the problem of bee sex determination and the lethal allele series of genes that causes about 50% of bee eggs to not hatch or to be eaten by the bees.* He would frequently have combs which contained eggs which he wanted to be egg free so he could use them for a test of a queen. I told him what I had found out about eggs being killed by sunlight and after that he would routinely expose combs to sunlight for several minutes to kill the eggs.

The dead eggs are rapidly eaten by the worker bees and a queen will almost immediately lay new eggs in those combs. Most beekeepers do not look at their brood combs every day or two and if they do they don't mark the particular cells to see if what they have done kills the young larvae or eggs. Well you might try it if you want something extra to do with your bees.

The procedure I used, which you might want to think about, was to place several hundred eggs in the incubator at one time and not to worry about the age of any one egg. Then I would look at the eggs before going home in the evening at about 16.30 or 17.00, then again the first thing in the morning a little past 08.00. Since the queen usually lays a bunch of eggs in a comb all about the same time, most of the eggs would be about the same age and after a time I would find that most eggs would hatch together and I would find the hatched eggs or young larvae ready for me to feed the first thing in the morning.

To get a lot of eggs to put in the incubator to work with is sometimes a problem but not always. I first tried - and it worked - slamming a comb containing many eggs down on a table top. Hundreds of eggs would be knocked out ready to work with. But that does not always work. Several other times I have tried this and the eggs would just not come out, and I expect there is a genetic component in egg stickiness - if that means anything. I suppose the simplest way of getting eggs is to cut out a sample of comb containing eggs, then shave the comb down so you can get at the eggs easily which is what Dr. DuPraw did when he was working on bee eggs. It is really not so difficult to pick up a bee egg and move it with the kind of needle used to transfer larvae for queen rearing and you can tell if you have injured the eggs a few hours after moving them by examining them under the 'scope.

After you have moved eggs and then hatched them, you are now ready to feed the little larvae. All you need is a hypodermic needle and syringe and some royal jelly. The royal jelly, of course, you can get from a queen cell and it keeps just fine most any place as long as it is sealed to prevent evaporation. Most people keep it refrigerated but this is not necessary even for long-term storage. What does happen is that crystals form in it that have to be broken up by stirring before you feed it to the little larvae.

The first person to my knowledge to feed bee larvae in the laboratory was Dr. Nevin Weaver who first did this way back forty years ago. He fed them every hour on the hour twenty-four hours a day. Dr. M.V. Smith, Ontario, Canada, later found out that he only had to feed the larvae once every twenty-four hours in order to rear them as adults. But he also found out that in order to do that, he had to maintain 95% RH around the larvae. Actually this is not so difficult to do in a home-made incubator as I have done it by placing four ice cube trays full of water in the incubator and allowing the fan to blow over them.

The larvae as they grow and develop turn circles as you will see when you feed them. They leave their tracks behind. When I was doing this almost all larvae fed that I started with, lived until they were about to turn into pre-pupae where in the bee hive they are sealed over by the bees. In a dish they defecate and create a real mess and in my experience many would then die. In nature inside a bee cell they also defecate but they all live.

Now the pre-pupae turn into pupae and finally into adults and a strange thing occurs, some turn into queens, some workers and some part queen and part worker. They all get the same food at the same time and on the same dish but they turn out differently. Everybody I know who has grown bee larvae in the incubator has had similar results and if you can make sense out of it I would like to hear from you.

Some rather smart folks have worked on this problem of what is called 'caste determination' and some people say they have solved the problem. I happen to disagree because if they know what to feed a larvae to make it a queen rather than a worker all people involved with queen rearing would add that material to their bees diet in order to make their bees rear better and better queens.

I hope you can see by now that there are lots of things you can do with your bees which are really interesting without all the hassle associated with producing a honey crop .

* Smith, M.V. 1959. Queen differentiation and the biological testing of royal jelly. N.Y., Ag. Exp. Sta. Mem. 356
   Weaver, N. 1956 Science 121:509
   Weaver, N. 1957 Ann. Entomol. Soc. Am. 50:283

BKQ 45
Spring 1996

Bee Larvae

Last BKQ my article was concerned with eggs and larvae and in this one the emphasis will be on larvae; what you can see if you look, and how they behave and grow. However, to do these described things with bees, you need a few things most beekeepers do not have but which would be extremely useful to those of you with an experimental mind, and to a bee club.

Of course I have my own ideas about what a club of hobby beekeepers should do and have, besides a meeting once a month and a real all-day session with live bees some time during the late spring or early summer. In addition, they should have a library of the eight or ten current important bee books and most of the important bee journals available; a platform scale to place under a hive to show periodic changes in weight, and a common place where members can put their hives of bees. In addition there are three other items which, while not of immediate necessity, really help a bee club to interest new members and keep older ones. These are, first, a compound microscope with magnifications from about 40X to 200X. Old 'scopes that are reasonably priced are excellent. The one I use is borrowed and must be seventy years old, but I can see nosema spores (*Nosema apis*), and internal parasitic mites (*Acarapis woodii*), easily with it. Plus you can easily see such things as a worker bee sting compared with a queen's, branched hairs, and many other things. Secondly, a dissecting microscope with magnifications of 7X to about 40X. I have taught many students artificial insemination, AI, of queen bees. Every now and then a student arrives who has never ever looked at a bee or part of any bee under magnification and they go nuts just looking for hours at things which have nothing to do with learning AI. They are looking at pollen baskets, branched hairs, antennae, tarsi and a million other things, instead of what they have paid me a lot of money to be looking at. These two implements complement each other and neither needs to be very expensive. I bought an expensive dissecting 'scope costing about $800 (£540) but I have seen very good second-hand ones for sale for $200 or less. The last item is an incubator which you can make from an old refrigerator. I bought a chicken egg incubator and removed the thermostat and heating element which were wired into the old 'fridge. Then I added a small circulating fan. The whole thing cost about $100 and works beautifully. I use it mostly to put queen cells to hold them until they emerge and my small rebuilt 'fridge can easily take a thousand cells.

Now with these items we can begin. As mentioned last BKQ you can watch an egg hatch and see what a fascinating process this is. And you can design all sorts of interesting experiments using bee eggs to see if they live and hatch. If you are of a scientific mind and want to know what is known about bee eggs, I will tell you, "Not much". One thing you will find right away is that you need to place pans of water or wet towels in the incubator or you will see the eggs have shrunk and died from dryness.

Now, if you have your incubator running at a correct humidity and temperature (95% RH and 34 'C) to hatch bee eggs, you will also have bee larvae which you can watch and feed or transfer back into the hive or use to raise queens with. When I was doing this years ago, I placed the bee eggs and larvae on clean beeswax, the kind that has just recently been melted, and never had any trouble hatching the eggs and feeding the larvae.

BKQ 57
Spring 1999

Drones

The vast majority of beekeepers don't like drones and rip out the drone brood when they find it in their hives. But the drones, the presence or absence of them, are an indicator and what they indicate or tell you about the hive's condition is what this article is about. The presence or absence of drone brood in the hive is used as an indicator of food sufficiency. Of course, there are exceptions to this generalization, so first we will talk about the exceptions.

The most important exception occurs in late summer or early fall in the north, when the first cold snap comes along with a killing frost. At that time of year there are plenty of stores in the hive so bees are not starving. But something tells the bees at that time, and I have no idea what it is, they no longer need drones. The adults are expelled from the hives and the drone eggs, larvae and pupae are eaten. Some books tell of this event as the "slaughter of the drones". In the southern part of the country, where I live, we do not get this drastic event. At least, I have never seen it because the bees don't have any drones to speak of at that time of year in their hives, anyway.

But in states like Wisconsin and New York, where I have worked, it is really dramatic. One day in January while I was a student at the University of Wisconsin, I was having an argument with Dr Farrar about this - the fact that bees killed off all drones in their hives in the fall. Farrar said, "No, some remain". We kept arguing and he had other work to do, so he flipped me the keys of the truck and told me to go look for myself.

There was snow on the ground and the temperature was about 15 F and the wind was blowing. I will never know how an 'ol Southern boy like me was able to survive four winters in Wisconsin. Man, it gets cold there! So here I was driving out to a bee yard all prepared to pry open a colony of bees just to prove it did not have drones.

At that time Farrar wintered all his colonies 3 high in full depth Langstroth boxes with the top box packed solid with honey. There were no bees in the top box. After a struggle I was able to get it loose from its frozen position and lifted it off where I encountered the bees.

When it is this cold the bees cannot move; all they do is stick their rear ends up at you, pointing their stingers right smack at the closest part of you. Some of you should try opening a hive on a really cold winter day. Try to pry the centre comb out and up, with that cluster of bees all stuck together like glue. It is a very difficult task with 10 frames or combs in a 10 frame box. Farrar always kept 9 in each box - as I do. This way, when a comb is removed, you don't roll the bees. It tends to get them very angry.

The centre comb and the centre of the cluster was comb 5 - and I succeeded with a great effort in getting it loose and pulling it free. The bees inside the cluster are warm and as soon as the comb is pulled up about 2 or 3 inches, those warm bees fly straight at you backward with their stings stuck straight out. Bee men are born to suffer; otherwise why be a beekeeper?

And yes, there were drones. 30 or 40 of them. And a small patch of worker brood too. That was a problem I always had with Farrar. He always knew exactly what was in every bee hive all the time. And, I will tell you, that is exasperating. He had always said "Queens will begin to lay eggs about the first of the year and will continue if they don't run out of pollen". Why, why, why would bees start rearing brood this far from spring and with no blooming flowers to provide them with food?

Where I have worked in southern parts of the US there is almost always something blooming all winter. On warm days, that is, warm enough for bees to fly, you will see them bringing in pollen. You should expect to see brood, that is small amounts, at all times of the year. But in Wisconsin? No way! ... But there it was!

The other exception to bees rearing drones is when the colony has a failing queen or has been queenless for a time and some of the worker bees are laying eggs. By the time you have kept 4 or 5 colonies for 10 years or so you will most likely have seen both of these rather peculiar events.

But let's get back to the main theme of using drones as an indicator. What does the presence of drones, adult and immature tell you? Remember how the bees are behaving and what they are doing in the mid-spring period until the midsummer period? During that part of the bee season when you open a hive and look in the brood nest you usually see every drone cell occupied with an immature drone, a pupa or larva. Because most beekeepers remove combs containing drone comb, the bees in their desperation build comb between brood chambers and fill it with drone brood.

If you have a number of colonies, say 20 or more, and have kept bees for 5 or 6 years, you most likely have experienced a period of time in the late spring when all of the bees look like a million. You think you are going to have the best season ever. But then you get hit by a cold, rainy period that seems to never end. You know the bees don't have much stores to tide them over and you don't want to feed sugar because as soon as the sun comes out your bees will be in a major honey flow. Finally, you look, and you see the worst thing that can happen, all your bees are starving. But now look a bit longer, you will see no drones, either adult or mature. The bees have eaten or killed them.

Years ago, when I was reading about bees - trying to learn

something, the author commented that the reason the bees raised so many drones was to use them as surplus food. So if you really notice what is going on in your hive with the drone brood, the young drone brood will sometimes disappear over-night. Think about that for a moment - a fully developed drone larva has almost enough nutrition to feed two worker larvae! Drone pupae, the bees can't eat these very well, so they throw them out the entrance; and some of you have seen that.

The spring buildup period is very critical for your bees; and if you are expecting and hoping to have your bees make a big crop of honey, you can't get that by subjecting them to starvation 15 minutes before the honey flow starts.

As a postscript; another very important indicator drones are good for is to capture the varroa mite. There are two parts to this, the first is as biological varroa control. This technique was developed by some European scientists years ago and simply involves placing full sheets of drone brood in the centre of the brood nest. When the drone brood has pupated, the comb is removed and discarded and another one placed from where it was removed. This technique is successful because the mites prefer and actively seek out drone brood to reproduce on. No chemicals are used and the mite populations are reduced.

The second part is to use drone pupae to determine the mite infestation rate. The drone pupae are uncapped and removed from their cells and the percent of infested larvae are determined. If you count 50% or more pupae infested, the colony should be treated with Apistan or Bayvarol strips. However, if a honey flow is in progress, wait to treat until you have removed your honey crop.

Drones Again
Sex in the Raw

I know from long experience that few people have read or studied the book written years ago by R.E. Snodgrass or read and studied the chapter co-authored with Dr Erickson, in the Hive and the Honey Bee, edition 1992, and I refer you specifically to pages 165-166. What I describe in this article can almost be done with the naked eye. A little magnification will help, but it is not necessary. In fact, I suggest that bee clubs could take this up in one of their meetings to learn more about sex, the sex of bees and insects in general. All you need for this exercise is a bunch of caught, flying drones, two or three for each person, and some tweezers or forceps, and a razor. A new cardboard cutter blade should be satisfactory. It will also be desirable to have a compound microscope available in order to see the movement of drone sperm. As you may be aware by now this article is sexually explicit.

OK, let's begin. Hold the drone securely with thumb and forefinger (index finger) of the left hand (if you are right-handed). Cut off the wings - you can use cuticle scissors if you wish, but I usually use a razor. Then with the razor make a slice on each side of the abdomen and a third slice across the top, right near the joining of the abdomen with the thorax. That joint is called a 'petiole'.

With the forceps, at the point where you cut the cuticle (the cuticle is the skin of an insect) near the thorax, lift the flap and pull it backward toward the rear of the drone and away from the thorax. Lay the drone over the middle finger, hold the flap down with the thumb and the drone's head with the index finger.

What you will see looks like some gooey mess, it is the digestive system of the drone. Take the forceps and remove all that mess which includes all the same organs found in the worker bee except the stomach. You will now see the reproductive organs of the drone and you should notice that it represents almost 75% of the contents of the drone's abdomen.

In doing this operation you will have to work on a live drone. If you kill the drone before doing this work you will make the same mistake as Snodgrass made which I will explain in a minute.

You should see two cream coloured glands laying on top of pearly white, much larger glands. The pearly-white, huge glands are the mucus glands and the yellowish or cream coloured glands on top of them are the seminal vesicles. The drone's testes are attached to the seminal vesicles and are difficult to see on old drones as they have degenerated. Now, with care (and a bit of skill) , use the forceps, right where the four glands join, grasp and pull them up and out. Watch carefully now and you will see the ejaculatory duct going to the penis bulb which looks like an enlargement and as you continue to pull it slowly out you will notice it is firmly anchored at the far end of the drone.

Now we need to have a bit of discussion. It has always amazed me to see the great similarities with the reproduction system of me, a man, a mammal, and a drone, an insect. Man has two testes, two seminal vesicles and an ejaculatory duct. Both the drone and man, when they ejaculate, not only ejaculate sperm, but most of the ejaculate is mucus from their respective mucus glands.

The biggest difference is that man continues to produce sperm all the time, but with the drone it is a one time thing. When the drone emerges from the pupal state, the sperm are all immature and are in the testes. If a dissection is made on a five-day-old drone, or one that cannot fly, you will see the testes are large and green-coloured and that at that time all the immature sperm are contained in the testes.

At about twelve days of age, all the sperm have matured and migrated into the seminal vesicles where they remain until the drone mates with a queen on their mating flight when they are expelled into the queen.

The mistake Snodgrass made, which you should avoid, was to kill the drone first, before making the dissection. The result of this is to cause all sperm to migrate into the penis bulb. He therefore wrote that the penis bulb was the storage place used by the drone to put his sperm before mating. Using a live drone you will see that all the sperm are in the seminal vesicles. Now is the time to get the compound microscope out with a slide. But first you will have to make up what is called a "normal saline solution". This is very easy to prepare. Measure out a cup of distilled water and dump a batch of table salt in it. Agitate it for a while, perhaps an hour, in order to dissolve as much salt as possible. Be sure some salt on the bottom is not dissolved. Pour this saturated salt solution into the cup and add it to exactly the same volume of water. With no measuring tools you have made a "normal saline solution". Actually it is just very close to being correct, but the error is not significant for these experiments.

Place a drop of this solution on a slide and place one of the seminal vesicles in the droplet. Cover with a cover slip (that is a very thin piece of glass to spread out the material you wish to see under the microscope). Push it down until the contents of the seminal vesicle are pushed out and then take a look through the lens. You will see this amazing sight of thousands and thousands of bee sperm swimming and swirling around. No need to hurry, they will live like that for an hour or two with no problem.

When you are out in the bee yard and drones are flying, catch one and pinch his head off. Usually this will cause a "partial eversion" of the drone. (See the diagram in Hive and the Honey Bee, page 166, Diagram "B".) The partial eversion causes the sperm to instantly move, because of an electrical impulse, from the seminal vesicles to the penis bulb. Now, if the drone's abdomen is squeezed gently from the part next to the thorax toward the rear, the ejaculation is complete and the sperm and mucus will form a little ball on the tip of the penis. The mucus is white and the sperm are dark yellow.

Easy to see even when wearing a bee veil!

Another exercise which can be done is to use a queen that you want to dispose of. This can be done in the bee yard even when wearing a veil. The exercise is to remove the queen's spermetheca (just using your propolis covered fingers), but the queen is killed first.

With thumb and forefinger grasp the tip end of the queen's abdomen, which will usually be the sting and the last two segments. Pull it apart from the rest of the queen. Lay this on the top of your thumb nail and gently push and roll the mess until you see a little white ball appear. The little ball is the queen's spermatheca.

At this point the little ball has a covering of very fine trachea which supplies oxygen to the sperm inside the ball. Rub the ball gently between thumb and forefinger until you see this white material come off the ball. If the ball is clear the queen has not been mated, if the ball appears cloudy it contains a few sperm and if it is the colour of the sperm you have already seen, she has a full complement of sperm.

A full complement of sperm in the spermetheca is between 5 and 7 million. (I have counted these several times using a haemocytometer, which is used to count blood cells). Sperm count from a drone will usually be about 10 million. Since we know a queen will mate with from 10 - 20 drones you see that a great number are lost somewhere in the mating process.

Here are a few facts and figures from the work of Dr O Mackenson who was one of the best scientists at artificial insemination I have ever observed. A microlitre is one millionth of a litre. A drone's ejaculate contains about one microlitre. An injection of 2.5 microlitres into the queen's oviduct resulted in an average of 2.97 million sperm in the spermetheca. Two injections given a day apart resulted in 4.11 million sperm and four injections of the same amount resulted in 5.52 sperms in the spermetheca. Counts of naturally-mated queens averaged 5.73 million sperm cells in the spermetheca.

I hope you find doing this exercise an interesting experience and understand what sex in the raw is all about.

Mackeson, O. and W. C. Roberts. 1948. A manual for the artificial insemination of queen bees. USDA ARS Blllietin ET-250.

Snodgrass, R. E. 1956 Anatomy or the honey bee. Cornell University Press, Ithaca, NY

Snodgrass, R.E. and LI-I. Erickson. The anatomy of the honey bee' in The Hive and the Honey Bee. 1992. Edited by], Graham, Dadant and Sons, Hamiliton, Illinois.

BKQ 52
Winter 1998

Naive Bees

Once upon a time, a long time ago, when I worked for the USDA bee lab at Baton Rouge I dreamed up an experiment which was rather unusual but didn't come out as it was supposed to, but was an awful lot of fun to figure out how to do. This article is about that fiasco (fiasco can have two meanings, it can mean a long-necked wine bottle, or a ridiculous failure: I refer to the latter). In some way perhaps this experiment was not a failure, but I know some will say it was quite ridiculous.

I had become acquainted with a professor of psychology at Louisiana State University and we got to talking about bees and behaviour problems and we both saw very quickly that we had a lot in common. The only real difference was that I was interested in bees and he was interested in people. Come to think of it there isn't much difference is there? (If you think there is try and figure out why your boss is a nutcase and why the queen bee won't do like, or as, she is supposed to do). He posed the question to me about learning in bees which went something like this: "Do the older bees teach the younger bees the language that Prof. K. von Frisch described so beautifully? Or do you suppose that the bees have all of this as an innate behaviour mechanism?".

Well, if you think about that for a moment or two, you must realise that never in nature do bees have to start from nothing - that is, there are always other bees in the hive when they emerge from their cells and there is almost always a queen too. But suppose you could arrange a hive with bees that had all emerged in an incubator and never had the opportunity to talk to another bee. Actually, that is not a very hard thing to arrange, so I did.

First, of course, I had to get all the important work out of the way which I was hired to do, which at that time was helping Dr O. Mackensen raise his queens and drones, and take care of his nucs and his artificially inseminated queens. That didn't really take up much of my time and there were lots of bee colonies available for me to work with and do whatever I wanted to. So, that being the case, I began arranging combs of brood above a queen excluder in a number of colonies so I would have sealed brood of a known age which, a few days before emerging, could be placed in an incubator to emerge and to provide me with 'naive' bees. Then I could watch their learning - see if they tumbled about, or any other way they might differ from ordinary bees.

We now have to digress a moment and define exactly what the term 'naive bees' means. The dictionary gives this meaning; "this is the result merely of a lack of experience". Then, extending the definition a bit further, it means, as applied to bees that I write about here, as those bees which have never ever had the experience of being with adult bees which have been exposed to the bee language hypothesis written about by Frisch. Nor have they ever been exposed to any other bees that could "talk" to them and give them "instructions" as to feeding the queen, or taking care of the eggs and larvae, and so on. My task then was to discover if bees learn everything needed to operate a hive properly all by themselves or do they have to have older, wiser and more knowledgeable bees to instruct them.

Two full days of emergence produced about two pounds of bees which I dumped or brushed into a box containing a comb of sealed brood and another of unsealed brood and another with some honey and water. I couldn't put this out in my regular bee yard because some strange bee might wander in and tell my naive bees everything, so I took them home and put them in my back yard, closed the entrance down to a very small place just big enough for one bee and waited patiently.

The next day at the entrance I saw some bee wings and bee legs, one or two of each, so I removed the cover and saw, to my surprise and shame, that all the bees were gone. The bottom board was covered with many wings and legs and the hive was overrun with large black wood ants which had invaded and destroyed my 'naive' colony. Well, as the saying goes, back to the drawing board.

After a few days contemplating this fiasco I realised that I had made a fundamental mistake, because I was setting up a bee colony from scratch but it obviously needed a queen. Then I thought, I can't just take any mated queen and put it in the hive as perhaps the queen would begin to issue royal commands and instruct all these 'naive' little things how to dance and find nectar and pollen and everything.

This did create a bit of a new problem which I had never done before, but which I knew was feasible. The new plan was to emerge the virgin queen in the incubator, give her a batch of day old 'naive' bees to look after her while she ages to the point where I can AI her and at that point have the frames of brood ready for the incubator. A pound or two of bees would emerge in the incubator which I could then unite with the inseminated queen. I had all the necessary equipment ready and I knew that I could do this experiment with no problem so I proceeded to set it into operation.

The queen emerged in the incubator as planned and I gave her a few bees for company and she was inseminated a few days later as planned. The queen and bees were kept in the incubator away from all other bees so that she could fully mature and have her ovaries ready to begin laying eggs as soon as the 'naive' bees in the hive fed and cared for her.

Combs of sealed brood were prepared and placed in the incubator, as before, and this time I decided to place this little colony in one of the LSU livestock pastures which is near the present location of the USDA ARS Bee Lab on Ben Hur Road. This location was miles from any live colonies of bees and also miles away from those ants which had destroyed my first attempt.

On Day 1, I went there with a small stool to sit near the entrance and watch. But first I opened the cover and looked inside and all looked okay. They had the appearance of a package of bees which had just been installed except that there were no bees coming from the entrance. I did the same thing on Day 2, and occasionally a bee or two would show its head at the entrance, then withdraw into the interior.

On Day 3, I took some sugar and water in a few dishes to see if I could begin training the bees to come to the sugar water. At this point I knew that the oldest bees were 4 days old and, according to many observations bees do not forage until they are ten days old. Very slowly I was able to get the bees to take this sugar water and as I moved the dish further and further from the entrance, the bees followed and were bringing more of their hive mates with them, but there weren't many, two or three I suppose, so I quit and went home for lunch leaving the few trained bees to forage from the dish which was now about 100 yards away from the hive.

As this was a beautiful warm Sunday afternoon, I didn't hurry back to my observation post, but when I did, a good two hours later, guess what? Yes, at least 50 bees were collecting from the dish, so I knew that my experiment was over.

So, what was I to do next? Well, I brought my stool to sit near the dish with the collecting bees, watched for a moment, and then idly caught and killed a few of the bees. Then a strange thing happened. I had not killed more than 4 or 5 bees when all of a sudden there were no bees at the feeding station. I knew exactly what had happened. The bees I picked up and killed had released a pheromone which instructed the other bees to go back home and I knew this pheromone was produced by the glands in the bees' head.

So my experiment was not a total fiasco. I learned that first, bees of five days or younger could be induced to forage - and perhaps that they actually do that in nature. Secondly, I learned that young emerging 'naive' bees do not need instructions from older and wiser sisters. Thirdly, I found out that bees have a repellent pheromone in their heads - but what is it and where exactly is the gland that produces it? Ah-ha! Problems, problems!

SECTION 5.
HONEY

BKQ 61
Summer 2000

Cream your Honey

The production of creamed honey on a small scale is within the bounds of all beekeepers without it being necessary to buy any extra equipment. By using this method you can provide your family, friends or customers with a creamy honey of a reliable and smooth consistency - ideal for any table!

When I lived in Louisiana, from 1950 to 1965, and worked at the Southern Bee States Culture Laboratory (as it was called then) I creamed honey almost every year. I never sold any, but I did give some to my friends and relatives. I don't know why more beekeepers don't do the same because it makes a very nice package to sell or to use.

These directions I am presenting now are the ones I used then. They require no new equipment and so there is no cost involved. It does require three conditions which are usually obtainable:

- The correct temperature conditions to start the process.

- A reliable method of heating the honey to a higher temperature.

- A source of good 'seed' honey.

1. Initiating the Process.

The correct temperature to begin the process needs to be no colder than 50 F (10 C) and no warmer than 60 F (15.5 C). If you have a cellar, check the temperature, it might be perfect. In the southern states like here in South Carolina, put an old freezer or fridge in the barn, garage, carport, or in a shaded place like the north side of a building. Put a thermometer in to check the temperature. When I lived in Louisiana (and Arizona) during most fall, winter and spring days, the temperature in such a chest will register between these two extremes.

If you do not have a way of controlling the temperature and have a cold snap or a heat wave, take the honey out of the chest and put it into your ordinary (connected) freezer. The freezer stops everything from happening and when the temperature in the old chest gets back to the correct range the honey can be put back again.

The ideal temperature for the honey to be creamed is 57 F (13.8 C) when the moisture content of the honey is at 18%. Honey is a viscous material. The colder it gets the harder it is for the sugar molecules to arrange themselves in a crystal lattice. And, if the temperature is too warm, the honey solution will take longer to solidify and the crystals will be larger and not fine. The whole object of the process is to turn out a very finely crystallised honey so that when you put in your mouth on your tongue and then placed on the roof of your mouth, you do not feel any granulation effect. If you do this test and it feels like sand, do it again.

2. Heating the Honey

Next, you need to heat the honey and it is critical that you do not get the honey too hot. It would be better if the honey could be heated in an apparatus similar to a double boiler but, whatever method you use, stir the honey constantly so that hot spots do not develop in the honey. My suggestion is that the honey is heated to about 130 F (54.4 C), and no hotter. Stick your finger in it to judge the temperature - if it is 130 F or over it will burn your finger and it will be too hot. Next, rapidly cool the honey to about 100 F (37.7 C). And it is at this temperature that the 'seed' honey is added.

3. 'Seeding' the Honey

I have always obtained my seed from other creamed honey with good results. However, the official instructions from Dr Dyce says that pushing granulated honey through a meat grinder does a very good job of getting starter seed.

When starting the process, I start with a one pound jar of clean, liquid honey and when the temperature of the honey to be creamed reaches about 100 F, I stir in a tablespoon of the starter seed until it is thoroughly mixed with the liquid. Place the jar of seeded honey into the cooler at 57 F and wait 2 - 3 days. After this time remove the honey and using a strong clean utensil stir the honey again. I use a 3/4" (29mm) thick piece of wooden dowel - you will see why it needs to be strong as the honey will be very stiff. Stir the honey until it is in a fluid state and uniformly so. Re-place the jar in the cooler for another 2 - 3 days and the honey will then be ready to eat.

Of course, you can use this as a starter to the next batch - say a five pound (2.2 kg) pail. About half a jar of the creamed honey will be needed to seed the 5 lb pail. If you wish to sell some of this honey, it is better (after the second stirring) to pour the honey into an opaque or waxed carton. Apparently the slightly mottled appearance of the creamed honey does not look very attractive through clear glass.

The process developed and patented in 1935 by Dr Dyce was offered to the honey industry under a small royalty fee. The amount of money collected was used primarily, I understand, to build the Dyce Honeybee Laboratory at Cornell University in New York State. The patent on this process has long since run out and you are, therefore, free to put such processed honey on the market.

What I have brought to your attention here is for the processing of relatively small quantities of honey for you or your friends' tables. For large scale commercial processing the principles are exactly the same - though I would advise you to visit a commercial producer or packer to see how it is done at that level.

Whether you are producing tons of creamed honey or just a few jars - REMEMBER: it is the temperature which is the critical factor.

References:

40th Edition of ABC and XYZ of Bee Culture. 1990. A I Root Co. Pages 143-146.

1992 Edition of The Hive and the Honey Bee. Dadant and Sons. Chapter 15, by Dr J Tew, pages 699 -703.

Photo Caption:

Creamed honey - easy to spread and smooth on the palate. It is best packed in an opaque or waxed card container as some customers might not like the mottled appearance. (Photos: Steve Taber)

www.ingramcontent.com/pod-product-compliance
Ingram Content Group UK Ltd.
Pitfield, Milton Keynes, MK11 3LW, UK
UKHW051853200426
11947UKWH00046B/1661

9 781908 904881